Science and the Unseen World

Arthur Stanley Eddington

Swarthmore Lecture, 1929

QUAKERbooks

First published 1929 by George Allen & Unwin
This edition published by Quaker Books, April
2007

www.quaker.org.uk

ISBN 978 0 901689 81 8

Cover image: photo of the total solar eclipse taken
on 21 June 2001 from near Lusaka Airport,
Zambia, by Aadil Desai of the Amateur
Astronomers' Association (Bombay).

Cover design: Hoop Associates

The Swarthmore Lecture Committee can be
contacted via the Clerk, c/o Woodbrooke Quaker
Study Centre, 1046 Bristol Road, Selly Oak,
Birmingham B29 6LJ.

The Swarthmore Lecture

The Swarthmore Lectureship was established by the Woodbrooke Extension Committee at a meeting held 9 December 1907: the minute of the Committee providing for an "annual lecture on some subject relating to the message and work of the Society of Friends". The name Swarthmore was chosen in memory of the home of Margaret Fox, which was always open to the earnest seeker after Truth, and from which loving words of sympathy and substantial material help were sent to fellow workers.

The Lectureship continues to be under the care of Woodbrooke Quaker Study Centre trustees, and is a significant part of the education work undertaken at and from Woodbrooke. The lectureship has a twofold purpose: first, to interpret to the members of the Society of Friends their message and mission; and second, to bring before the public the spirit, aims and fundamental principles of Friends. The lecturers alone are responsible for any opinions expressed.

The lectureship provides both for the publication of a book and for the delivery of a lecture, the latter usually at the time of Britain Yearly Meeting of the Society of Friends (London Yearly Meeting up to 1994). The present Lecture was delivered at Friends House, London, on the evening preceding the Yearly Meeting, 1929.

Woodbrooke
Quaker Study Centre

The Author

Arthur Stanley Eddington was born in 1882. He took many awards as a student in sciences, although he also considered an arts career. After working at the Royal Observatory in Greenwich, in 1913 he became Plumian Professor of Astronomy, University of Cambridge, then in 1914 director of the university's observatory, and a fellow of the Royal Society. He played an essential role in testing Einstein's theory of relativity, at an eclipse in 1919. Einstein considered him his best interpreter.

A lifelong Quaker, he was a conscientious objector in the 1914–18 war, but the university separately applied for his exemption as "indispensable". He was chairman of the National Peace Council 1941–43, and was a sponsor of the Peace Pledge Union. He died in 1944.

His other books include *The Nature of the Physical World* (1928) and *The Expanding Universe* (1933).

Contents

Foreword

Arthur Eddington was one of the greatest Quaker scientists, being the pioneer developer of astrophysics (the application of physical theory to understanding how stars function), and one of the earliest proponents of general relativity theory (Einstein's theory of gravitation). This lecture was given in 1929, the year Hubble provided the observational evidence for the expansion of the universe but before the concept of the expanding universe was generally known and accepted, and when quantum theory was only newly developed. He presents us with a sound physical and astrophysical view from that time. You might be misled from your present knowledge to read back into the past and think this talk presents the idea of the expanding universe, but it does not do so. This is an example of the progress of science: the basics remain the same, but we have learnt a huge amount more about physics and the universe since then. Nevertheless the elements of what Eddington presents us in terms of the relation of science to metaphysics remain unaltered.

He puts a strong line against simplistic reductionism in relation to our minds, and gives us a timely reminder: "In comparing the certainty of things spiritual and things temporal, let us not forget this – Mind is the first and most direct thing in experience; all else is remote inference". This is a good antidote to the many who confuse their models of reality with reality itself. He emphasises that when we ask the question, "What are we to think of it all? What is it all about?", the answer must embrace but not be limited to the scientific answer. His lecture explores this in a delightful way, that remains fully relevant today. He comments, "Dismiss the idea that natural law may swallow up religion; it cannot even tackle the multiplication table single-handed." His penetrating analysis is as valuable today as when it was delivered. He illustrates clearly how the Quaker position is fully consonant with a scientific outlook.

George Ellis, Cape Town

April 2007

George Ellis is Distinguished Professor of Complex Systems at the University of Cape Town, and holds a Visiting Chair in Astronomy at Queen Mary College in London. In 2004 he was awarded the Templeton Prize for discoveries about spiritual realities. His pamphlet *Science in Faith and Hope: an interaction* is also published by Quaker Books.

I

Outline of evolution leading to the advent of Man in the physical world

Looking back through the long past we picture the beginning of the world – a primeval chaos which time has fashioned into the universe that we know. Its vastness appals the mind; space boundless though not infinite, according to the strange doctrine of science. The world was without form and almost void. But at the earliest stage we can contemplate the void is sparsely broken by tiny electric particles, the germs of the things that are to be; positive and negative they wander aimlessly in solitude, rarely coming near enough to seek or shun one another. They range everywhere so that all space is filled, and yet so empty that in comparison the most highly exhausted vacuum on earth is a jostling throng. In the beginning was vastness, solitude and the deepest night. Darkness was upon the face of the deep, for as yet there was no light.

The years rolled by, million after million. Slight aggregations occurring casually in one place and another drew to themselves more and more particles. They warred for sovereignty, won and lost their spoil, until the matter was collected round centres of condensation leaving vast empty spaces from which it had ebbed away. Thus gravitation slowly parted the primeval chaos. These first divisions were not the stars but what we should call "island universes" each ultimately to be a system of some thousands of millions of stars. From our own island universe we can discern the other islands as spiral nebulae lying one beyond another as far as the telescope can fathom. The nearest of them is such that light takes 900,000 years to cross the gulf between us. They acquired rotation (we do not yet understand how) which bulged them into flattened form and made them wreathe themselves in spirals. Their forms, diverse yet with underlying regularity, make a fascinating spectacle for telescopic study.

As it had divided the original chaos, so gravitation subdivided the island universes. First the star clusters, then the stars themselves were separated. And with the stars came light, born of the fiercer turmoil which ensued when the electrical particles were drawn from their solitude into dense throngs. A star is not lump of matter casually thrown together in the general confusion; it is of nicely graded size. There is relatively not much more diversity in the masses of new-born stars than in the masses of new-born babies. Aggregations rather greater than our Sun have a strong tendency to subdivide, but when the mass is reduced a little the danger quickly passes and the impulse to subdivision is satisfied. Here it would seem the work of creation might cease. Having carved chaos into stars, the first evolutionary impulse has reached its goal. For many billions of years the stars may continue to shed their light and heat through the world, feeding on their own matter which disappears bit by bit into ætherial waves.

Not infrequently a star, spinning too fast or strained by the radiant heat imprisoned within it, may divide into two nearly equal stars, which remain yoked together as a double star; apart from this no regular plan of further development is known. For what might be called the second day of creation we turn from the general rule to the exceptions. Amid so many myriads there will be a few which by some rare accident have a fate unlike the rest. In the vast expanse of the heavens, the traffic is so thin that a star may reasonably count on travelling for the whole of its long life without serious risk of collision. The risk is negligible for any individual star; but ten thousand million stars in our own system and more in the systems beyond afford a wide playground for chance. If the risk is one in a hundred millions some unlucky victims are doomed to play the role of "one." This rare accident must have happened to our Sun – an accident to the Sun but to us the cause of our being here. A star journeying through space casually overtook the Sun, not indeed colliding with it, but approaching so close as to raise a great tidal wave. By this disturbance jets of matter spurted out of the Sun; being carried round by their angular momentum they did not fall back again but condensed into small globes – the planets.

By this and similar events there appeared here
and there in the universe something outside
Nature's regular plan, namely a lump of matter
small enough and dense enough to be cool. A
temperature of ten million degrees or more
prevails through the greater part of the interior of
a star; it cannot be otherwise so long as matter
remains heaped in immense masses. Thus the
design of the first stage of evolution seems to been
that matter should ordinarily be endowed with
intense heat. Cool matter appears as an
afterthought. It is unlikely that the Sun is the only
one of the starry host to possess a system of
planets, but it is believed that such development is
very rare. In these exceptional formations Nature
has tried the experiment of finding what strange
effects may ensue if matter is released from its
usual temperature of millions of degrees and
permitted to be cool.

Out of the electric charges dispersed in the
primitive chaos ninety-two different kinds of
matter – ninety-two chemical elements – have
been built. This building is also a work of
evolution, but little or nothing is known as to its
history. In the matter which we handle daily we
find the original bricks fitted together and cannot
but infer that somewhere and somewhen a process
of matter-building has occurred. At high temperature this diversity of matter remains as it were
latent; little of consequence results from it. But in
the cool experimental stations of the universe the

differences assert themselves. At root the diversity of the ninety-two elements reflects the diversity of the integers from one to ninety-two; because the chemical characteristics of element No. 11 (sodium) arise from the fact that it has the power at low temperatures of gathering round it eleven negative electric particles; those of No. 12 (magnesium) from its power of gathering twelve particles; and so on.

It is tempting to linger over the development out of this fundamental beginning of the wonders studied in chemistry and physics, but we must hurry on. The provision of certain cool planetary globes was the second impulse of evolution, and it has exhausted itself in the formation of inorganic rocks and ores and other materials. We must look to a new exception or abnormality if anything further is to be achieved. We can scarcely call it an accident that among the integers there should happen to be the number 6; but I do not know how otherwise to express the fact that organic life would not have begun if Nature's arithmetic had overlooked the number 6. The general plan of ninety-two elements, each embodying in its structural pattern one of the first ninety-two numbers, contemplates a material world of considerable but limited diversity; but the element carbon, embodying the number 6, and because of the peculiarity of the number 6, rebels against limits. The carbon atoms love to string themselves in long chains such as those which give toughness to a soap-film. Whilst other atoms organise themselves in twos and threes or it may be in tens, carbon atoms organise themselves in hundreds and thousands. From this potentiality of carbon to form more and more elaborate structure a third impulse of evolution arises.

I cannot profess to say whether anything more than this prolific structure-building power of carbon is involved in the beginning of life. The story of evolution here passes into the domain of the biological sciences for which I cannot speak, and I am not ready to take sides in the controversy between the Mechanists and the Vitalists. So far as the earth is concerned the history of development of living forms extending over nearly a thousand million years is recorded (though with many breaks) in fossil remains. Looking back over the geological record it would seem that Nature made nearly every possible mistake before she reached her greatest achievement Man – or perhaps some would say her worst mistake of all. At one time she put her trust in armaments and gigantic size. Frozen in the rock is the evidence of her failures to provide a form fitted to endure and dominate – failures which we are only too ready to imitate. At last she tried a being of no great size, almost defenceless, defective in at least one of the more important sense-organs; one gift she bestowed to save him from threatened extinction – a certain stirring, a restlessness, in the organ called the brain.

And so we come to Man.

II

The questioning voice, "What doest thou here?"

It is with some such thoughts as these of the
relation of Man to the visible universe that the
scientifically minded among us approach the
problem of his relation to the Unseen World. It is
not with any dogmatic challenge that I have given
this outline of evolution. Part of what I have
described seems to be securely established; other
parts involve a considerable element of conjecture
– the best we can do to string together fragmentary
knowledge. Scientific theories have blundered in
the past; they blunder no doubt to-day; yet we
cannot doubt that along with the error there come
gleams of a truth for which the human mind is
impelled to strive. So brief a summary cannot
convey the true spirit and intention of this
scientific probing of the past, any more than the
spirit of history is conveyed by a table of dates. We
seek the truth; but if some voice told us that a few
years more would see the end of our journey, that
the clouds of uncertainty would be dispersed, and
that we should perceive the whole truth about the
physical universe, the tidings would be by no
means joyful. In science as in religion the truth
shines ahead as a beacon showing us the path; we
do not ask to attain it; it is better far that we be
permitted to seek.

I daresay that most of you are by no means reluctant to accept the scientific epic of the Creation, holding it perhaps as more to the glory of God than the traditional story. Perhaps you would prefer to tone down certain harshnesses of expression, to emphasise the forethought of the Creator in the events which I have called accidents. I would not venture to say that those who are eager to sanctify, as it were, the revelations of science by accepting them as new insight into the divine power are wrong. But this attitude is liable to grate a little on the scientific mind, forcing its free spirit of inquiry into one predetermined mode of expression; and I do not think that the harmonising of the scientific and the religious outlook on experience is assisted that way. Perhaps our feeling on this point can be explained by a comparison. A business man may believe that the hand of Providence is behind his commercial undertakings as it is behind all the vicissitudes of his life; but he would be aghast at the suggestion that Providence should be entered as an asset in his balance sheet. I think it is not irreligion but a tidiness of mind, which rebels against the idea of permeating scientific research with a religious implication.

Probably most astronomers, if they were to speak frankly, would confess to some chafing when they are reminded of the psalm "The heavens declare the glory of God." It is so often rubbed into us with implications far beyond the simple poetic thought awakened by the splendour of the star-clad sky. There is another passage from the Old Testament that comes nearer to my own sympathies – "And behold the Lord passed by, and a great and strong wind rent the mountains, and brake in pieces the rocks before the Lord; but the Lord was not in the wind: and after the wind an earthquake; but the Lord was not in the earthquake: and after the earthquake a fire; but the Lord was not in the fire: and after the fire a still small voice. . . And behold there came a voice unto him, and said, What doest thou here, Elijah?"

Wind, earthquake, fire – meteorology, seismology, physics – pass in review, as we have been reviewing the natural forces of evolution; the Lord was not in them. Afterwards, a stirring, an awakening in the organ of the brain, a voice which asks "What doest thou here?"

III

Changing views of the scope of physical theory and the ideal of physical explanation

We have busied ourselves with the processes by which the electric particles widely diffused in primeval chaos have come together to build the complexity of a human being; we cannot but acknowledge that a human being involves also something incommensurable with the kind of entities we have been treating of. I do not mean to say that consciousness has not undergone evolution; presumably its rudiments exist far down the scale of animal life. But it is a constituent or an aspect of reality which our survey of the material world leaves on one side. Hence arises insistently the problem of the dualism of spirit and matter. On the one side there is consciousness stirring with activity of thought and sensation; on the other side there is a material brain, a maelstrom of scurrying atoms and electric charges. Incommensurable as they are, there is some kind of overlap or contact between them. As the mind is traversed by a certain thought the atoms at some point of the brain range themselves so as to start a material impulse transmitting the mental command to a muscle; or again a nervous impulse arrives from the outer world, and as the atoms of a brain-cell move in response to the physical forces simultaneously a sensation of pain occurs in the mind.

Let us for a moment consider the most crudely materialistic view of this connection. It would be that the dance of atoms in the brain really constitutes the thought, that in our search for reality we should replace the thinking mind by a system of physical objects and forces, and that by so doing we strip away an illusory part of our experience and reveal the essential truth which it so strangely disguises. I do not know whether this view is still held to any extent in scientific circles, but I think it may be said that it is entirely out of keeping with recent changes of thought to the fundamental principles of physics. Its attractiveness belonged to a time when it was considered that the way to understand or explain a scientific phenomenon was to make a concrete mechanical model of it.

I cannot in a few moments make clear a change of thought which it has taken a generation to accomplish. I can only say that physical science has turned its back on all such models, regarding them now rather as a hindrance to the apprehension of the truth behind the phenomena. We have the same desire as of old to get to the bottom of things, but the ideal of what constitutes a scientific explanation has changed almost beyond recognition. And if today you ask a physicist what he has finally made out the æther or the electron to be, the answer will not be a description in terms of billiard balls or fly-wheels or anything concrete; he will point instead to a number of symbols and a set of mathematical equations which they satisfy. What do the symbols stand for? The mysterious reply is given that physics is indifferent to that; it has no means of probing beneath the symbolism. To understand the phenomena of the physical world it is necessary to know the equations which the symbols obey but not the nature of that which is being symbolised. It would be irrelevant here to defend this change, to make clear the intellectual satisfaction afforded by these symbolic equations, or to explain why the demand of the layman for a concrete explanation has to be set aside. We have, however, to see how this newer outlook has modified the challenge from the material to the spiritual world.

For those who were bent on finding a model for everything, the material brain appeared in the light of a ready-made model of the mind. And being a model, it was for them the full explanation of the mind. A mechanism of concrete particles, like the billiard-ball atoms of the brain, was their ideal of an explanation. They were hoping similarly to find a mechanism of gyrostats and cog-wheels to explain the æther. The cog-wheels of the æther were hidden, but the cog-wheels of the mind seemed to be at any rate partly exposed. The mere sight of such machinery gave them a feeling of satisfaction, even if they could not tell in the least how it worked. I am not here greatly concerned with the question whether, or to what extent, the brain-cells may rightly be regarded as the cog-wheels of the mind. What I wish to point out is that we no longer have the disposition which, as soon as it scents a piece of mechanism, exclaims "Here we are getting to bedrock. This is what things should resolve themselves into. This is ultimate reality." Physics to-day is not likely to be attracted by a type of explanation of the mind which it would scornfully reject for its own æther.

Perhaps the most essential change is that we are no longer tempted to condemn the spiritual aspects of our nature as illusory because of their lack of concreteness. We have travelled far from the standpoint which identifies the real with the concrete. Even the older philosophy found it necessary to admit exceptions; for example, time must be admitted to be real, although no one could attribute to it a concrete nature. Nowadays time might be taken as typical of the kind of stuff of which we imagine the physical world to be built. Physics has no direct concern with that feeling of "becoming" in our consciousness which we regard as inherently belonging to the nature of time, and it treats time merely as a symbol; but equally matter and all else that is in the physical world have been reduced to a shadowy symbolism.

We all share the strange delusion that a lump of matter is something whose general nature is easily comprehensible whereas the nature of the human spirit is unfathomable. But consider how our supposed acquaintance with the lump of matter is attained. Some influence emanating from it plays on the extremity of a nerve, starting a series of physical and chemical changes which are propagated along the nerve to a brain cell; there a mystery happens, and an image or sensation arises in the mind which cannot purport to resemble the stimulus which excites it. Everything known about the material world must in one way or another have been inferred from these stimuli transmitted along the nerves. It is an astonishing feat of deciphering that we should have been able to infer an orderly scheme of natural knowledge from such indirect communication. But clearly there is one kind of knowledge which cannot pass through such channels, namely knowledge of the intrinsic nature of that which lies at the far end of the line of communication. The inferred knowledge is a skeleton frame, the entities which build the frame being of undisclosed nature. For that reason they are described by symbols, as the symbol *x* in algebra stands for an unknown quantity.

The mind as a central receiving station reads the dots and dashes of the incoming nerve-signals. By frequent repetition of their call-signals the various transmitting stations of the outside world become familiar. We begin to feel quite a homely acquaintance with 2L0 and 5XX. But a broadcasting station is not *like* its call signal; there is no commensurability in their nature. So too the chairs and tables around us which broadcast to us incessantly those signals which affect our sight and touch cannot in their nature be like unto the signals or to the sensations which the signals awake at the end of their journey.

Penetrating as deeply as we can by the methods of physical investigation into the nature of a human being we reach only symbolic description. Far from attempting to dogmatise as to the nature of the reality thus symbolised, physics most strongly insists that its methods do not penetrate behind the symbolism. Surely then that mental and spiritual nature of ourselves, known in our minds by an intimate contact transcending the methods of physics, supplies just that interpretation of the symbols which science is admittedly unable to give. It is just because we have a real and not merely a symbolic knowledge of our own nature that our nature seems so mysterious; we reject as inadequate that merely symbolic description which is good enough for dealing with chairs and tables and physical agencies that affect us only by remote communication.

In comparing the certainty of things spiritual and things temporal, let us not forget this – Mind is the first and most direct thing in experience; all else is remote inference.

That environment of space and time and matter, of light and colour and concrete things, which seems so vividly real to us is probed deeply by every device of physical science and at bottom we reach symbols. Its substance has melted into shadow. None the less it remains a real world if there is a background to the symbols – an unknown quantity which the mathematical symbol x stands for. We think we are not wholly cut off from this background. It is to this background that our own personality and consciousness belong, and those spiritual aspects of our nature not to be described by any symbolism or at least not by symbolism of the numerical kind to which mathematical physics has hitherto restricted itself. Our story of evolution ended with a stirring in the brain-organ of the latest of Nature's experiments; but that stirring of consciousness transmutes the whole story and gives meaning to its symbolism. Symbolically it is the end, but looking behind the symbolism it is the beginning.

IV

Both a scientific and a mystical outlook are involved in the "problem of experience"

What is the problem that is contemplated when we discuss the possible conflict of the scientific and the religious outlook? I think that so far as the Society of Friends is concerned we should define it as the problem presented by experience – the problem of the proper orientation of our minds towards the different elements of our experience. If science claims in any way to be a guide to life it is because it deals with experience, or part of experience. And if religion is not an attitude towards experience, if it is just a creed postulating an ineffable being who has no contact with ourselves, it is not the kind of religion which our Society stands for. The interaction of ourselves with our environment is what makes up experience. Part of that interaction consists in the sensations associated with impulses coming through our sense-organs; it is by following up this element of experience that we reach the scientific problem of the physical world. But surely experience is broader than this, and the problem of experience is not limited to the interpretation of sense-impressions.

Picture first consciousness as a bundle of sense-impressions and nothing more. As the sensations succeed one another, as they are compared in one consciousness and another, from somewhere comes the query "What are we to think of it all? What is it all about?" To answer this is the purpose of science. But picture again consciousness, not this time as a bundle of sense-impressions but as we intimately know it, responsible, aspiring, yearning, doubting, originating in itself such impulses as those which urge the scientist on his quest for truth. "What are we to think of it all? What is it all about?" This time the answer must be broader, embracing but not limited to the scientific answer.

Normally it is my task to propagate the truths of science, to urge its outstanding importance, and to tread myself the way by which it seeks an understanding of the phenomena which we experience. It is far from my thought to disparage what we gain by this quest. As truly as the mystic, the scientist is following a light; and it is not a false or an inferior light. Moreover the answers given by science have a singular perfection, prized the more because of the long record of toil and achievement behind them. Why then do I not produce one of these scientific answers now? Simply because before giving an answer, it is usual to listen to the question that is put. It is no use having ready a flawless answer if people will not put to you the question it is intended for. So far as I can judge, the kind of question to which I have exposed myself by coming here to-night is, What is the proper orientation of a rational being towards that experience which he so mysteriously finds himself partaking of? What conception of his surroundings should guide him as he sets about the fulfilment of the life bestowed on him? Which of those strivings and feelings which make up his nature are to be nourished, and which rejected as the seed of illusion? The desire for truth so prominent in the quest of science, a reaching out of the spirit from its isolation to something beyond, a response to beauty in nature and art, an Inner Light of conviction and guidance – are these as much a part of our being as our sensitivity to sense-impressions? I have no ready-made answer for these questions. Study of the scientific world cannot prescribe the orientation of something which is excluded from the scientific world. The scientific answer is relevant so far as concerns the sense-impressions interlocked with the stirring of the spirit, which indeed form an important part of the mental content. For the rest the human spirit must turn to the unseen world to which it itself belongs.

Some would put the question in the form "Is the unseen world revealed by the mystical outlook a reality?" Reality is one of indeterminate words which might lead to infinite philosophical discussions and irrelevancies. There is less danger of misunderstanding if we put the question in the form: "Are we, in pursuing the mystical outlook, facing the hard facts of experience?" Surely we are. I think that those who would wish to take cognisance of nothing but the measurements of the scientific world made by our sense-organs are shirking one of the most immediate facts of experience, namely that consciousness is not wholly, nor even primarily a device for receiving sense impressions.

We may the more boldly insist that there is another outlook than the scientific one, because in practice a more transcendental outlook is almost universally admitted. I cannot do better than quote a memorable passage from the Swarthmore Lecture by J. S. Hoyland last year.

"There is an hour of the Indian night, a little before the first glimmer of dawn, when the stars are unbelievably clear and close above, shining with a radiance beyond our belief in this foggy land. The trees stand silent around one with a friendly presence. As yet there is no sound from awakening birds; but the whole world seems to be intent, alive, listening, eager. At such a moment the veil between the things that are seen and the things that are unseen becomes so thin as to interpose scarcely any barrier at all between the eternal beauty and truth and the soul which would comprehend them."

Here is an experience which the "observer" as technically defined in scientific theory knows nothing of. The measuring appliances which he reads declare that the stars are just as remote as they always have been, nor can he find any excuse in his measures for the mystic thought which has taken possession of the mind and dominated the sense-impressions. Yet who does not prize these moments that reveal to us the poetry of existence? We do not ask whether philosophy can justify such an outlook on nature. Rather our system of philosophy is itself on trial; it must stand or fall according as it is broad enough to find room for this experience as an element of life. The sense of values within us recognises that this is a test to be passed; it is as essential that our philosophy should survive this test as that it should survive the experimental tests supplied by science.

In the passage I have quoted there is no direct reference to religious mysticism. It describes an orientation towards nature accepted by religious and irreligious alike as proper to the human spirit – though not to the ideal "observer" whose judgments form the canon of scientific experience. The scientist who from time to time falls into such a mood does not feel guilty twinges as though he had lapsed in his devotion to truth; he would on the contrary feel deep concern if he found himself losing the power of entering into this kind of feeling. In short our environment may and should mean something towards us which is not to be measured with the tools of the physicist or described by the metrical symbols of the mathematician. We cannot argue that because natural mysticism is universally admitted in some degree therefore religious mysticism must necessarily be admitted; but objections to religious mysticism lose their force if they can equally be turned against natural mysticism. If we claim that the experience which comes to us in our silent meetings is one of the precious elements that make up the fullness of life, I do not see how science can gainsay us. Let it pause before rushing in to apply a supposed scientific test; for such a test would go much too far, stripping away from our lives not only our religion but all our feelings which do not belong to the function of a measuring-machine.

In justifying the place of religious experience in human life, we have not to consider it from the point of view of propagating a creed. We do not send missionaries to the blind to persuade them that it will be to their benefit to believe that a world of light and colour exists for other men gifted with eyes. We should not argue with the blind man who maintained that sight was an illusion to which some abnormal people were subject. Therefore in speaking of religious experience I do not attempt to prove the existence of religious experience, any more than in lecturing on optics I should attempt to prove the existence of sight. What I may attempt is to dispel the feeling that in using the eye of the body or the eye of the soul, and incorporating what is thereby revealed in our conception of reality, we are doing something irrational and disobeying the leading of truth which as scientists we are pledged to serve.

V

The irrelevancy of "natural law" to some aspects of mind and consciousness

I have already said that science is no longer disposed to identify reality with concreteness. Materialism in its literal sense is long since dead. But its place has been taken by other philosophies which represent a virtually equivalent outlook. The tendency today is not to reduce everything to manifestations of matter – since matter now has only a minor place in the physical world – but to reduce it to manifestations of the operation of natural law. By "natural law" is here meant laws of the type prevailing in geometry, mechanics, and physics which are found to have this common characteristic – that they are ultimately reducible to mathematical equations. They may also be defined by a less technical property, viz., they are laws which, unlike human law, are never broken. It is this belief in the universal dominance of scientific law which is nowadays generally meant by materialism.

The harmony and simplicity of scientific law appeals strongly to our aesthetic feeling. It illustrates one kind of perfection, such as we might perhaps think worthy to be associated with the mind of God. One of the important questions that we have to face is whether the unseen world is governed by a like scheme of law. I am aware that many religious writers have felt no objection to, and even welcomed, the intrusion of natural law into the spiritual domain. (Probably, however, they are using the term "natural law" in a more elastic sense than that in which the materialist understands it.) Why (they ask) should we insist for ourselves on exemption from a kind of government which as displayed in inorganic nature might be hailed as a manifestation of divine perfection? But I am sure that those who take this view have never understood and faced the meaning of the ideal scheme of scientific law. What they would welcome is not science but pseudo-science. Analogies can be drawn between spiritual and natural phenomena which may serve to press home a moral lesson. For example, one of Kirchoff's famous laws of radiation states that the absorbing power of substances is proportional to the emitting power, so that the best absorbers are also the best emitters. That might make a good text for a sermon. But if ever scientific law makes a serious inroad into the spiritual domain the consequences will not be limited to supplying texts for sermons.

Natural law is not applicable to the unseen world behind the symbols, because it is unadapted to anything except symbols, and its perfection is a perfection of symbolic linkage. You cannot apply such a scheme to the parts of our personality which are not measurable by symbols any more than you can extract the square root of a sonnet. There is a kind of unity between the material and the spiritual worlds – between the symbols and their background – but it is not the scheme of natural law which will provide the cement.

In saying this I am not forgetting the likelihood of great future developments of science which may and indeed must bring to light types of natural law of which as yet we have no conception. Thus I do not judge the problem of life (in so far as it can be dissociated from consciousness) to be impregnable to the attack of physics. It is a matter of keen controversy among biochemists whether physics and chemistry as they stand are adequate to deal with the properties of living organisms. I express no opinion; but, in any case, whether they are adequate or not today I cannot assume that future revolutions of science and the admission of new fundamental conceptions will not make them adequate. It is when life is associated with consciousness that we reach different ground altogether. To those who have any intimate acquaintance with the laws of chemistry and physics the suggestion that the spiritual world could be ruled by laws of allied character is as preposterous as the suggestion that a nation could be ruled by laws like the laws of grammar. The essential difference, which we meet in entering the realm of spirit and mind, seems to hang round the word "Ought."

This limitation of natural law to a special domain would be more obvious but for a confusion in our use of the word law. In human affairs it means a rule, fortified perhaps by incentives or penalties, which may be kept or broken. In science it means a rule which is never broken; we suppose that there is something in the constitution of things which makes its non-fulfilment an impossibility. Thus in the physical world what a body does and what a body ought to do are equivalent; but we are well aware of another domain where they are anything but equivalent. We cannot get away from this distinction. Even if religion and morality are dismissed as illusion, the word "Ought" still has sway. The laws of logic do not prescribe the way our minds think; they prescribe the way our minds ought to think.

Suppose we concede the most extravagant claims that might be made for natural law, so that we allow that the processes of the mind are governed by it; the effect of this concession is merely to emphasise the fact that the mind has an outlook which transcends the natural law by which it functions. If, for example, we admit that every thought in the mind is represented in the brain by a characteristic configuration of atoms, then if natural law determines the way in which the configurations of atoms succeed one another it will simultaneously determine the way in which thoughts succeed one another in the mind. Now the thought of "7 times 9" in a boy's mind is not seldom succeeded by the thought of "65". What has gone wrong? In the intervening moments of cogitation everything has proceeded by natural laws which are unbreakable. Nevertheless we insist that something has gone wrong. However closely we may associate thought with the physical machinery of the brain, the connection is dropped as irrelevant as soon as we consider the fundamental property of thought – that it may be correct or incorrect. The machinery cannot be anything but correct. We say that the brain which produces "7 times 9 are 63" is better than the brain which produces "7 times 9 are 65"; but it is not as a servant of natural law that it is better. Our approval of the first brain has no connection with natural law; it is determined by the type of thought which it produces, and that involves recognising a domain of the other type of law – laws which ought to be kept, but may be broken. Dismiss the idea that natural law may swallow up religion; it cannot even tackle the multiplication table single-handed.

VI

The importance of "significances" and the consequences of ruling them outside the scope of inquiry

Let me play the role of materialist philosopher a few moments longer. The electric particles in obedience to the laws of physics have come together and built human brains. Still in obedience to those laws, they have by their evolutions brought about and stored in those brains the thoughts that make up the sum of human knowledge. Those unbreakable laws have decreed that tonight some of that accumulated knowledge is to be unloosed on you in the form of a lecture. I must hope that you too will be good materialists and feel a due interest in the phenomenon that is proceeding, observing the curious effects of Maxwell's laws, the laws of thermodynamics and other physical causes that are leading to the emission of a modulated system of sound-waves. But no; I was forgetting. That is how as materialists you *ought* to think of my lecture; but "ought" is outside natural law. I cannot expect more than that your brains will react towards the lecture in accordance with the unbreakable laws which govern them; and those who happen to fall asleep may claim that it was decreed by those laws. This is, of course, a very old *reductio ad absurdum;* and he would be a very shallow materialist who has not appreciated the difficulty and persuaded himself that he has found an answer to it. I am not very curious as to how he surmounts the difficulty or whether his justification is valid. The upshot is that he connives at an attitude towards knowledge which does not treat it as something secreted in the brain by the operation of unbreakable laws of nature. It is to be judged in relation to its truth or untruth not in relation to any supposed theory of its origin.

Truth and untruth belong to the realm of significance and values. I am not able to agree entirely with the assertion commonly made by scientific philosophers that science, being solely concerned with correct and colourless description, has nothing to do with significances and values. If it were literally true, it would mean that, when the significance of our lives and of the universe around us is under discussion, science is altogether dumb. But there is this much truth in it. If we are to present science as a self-contained scheme, owing nothing to any judgments we may have formed by methods for which science does not take responsibility, then no doubt significances must be ruled outside its scope. This may be called the official attitude of science. Officially the scientist is just an adept at solving certain problems; he has no curiosity as to how these problems have come to be set; it is a complete surprise to him that mankind struggling after the eternal verities should take serious note of his pastime. But I think no one would venture to speak to a public audience on any scientific topic unless he were prepared to transgress beyond the official attitude. Imagine a speaker on evolution presenting a purely colourless description of the sequence of living forms and the struggle for existence, without ever hinting at an underlying significance for us of this change in our belief as to Man's place in nature.

The religious seeker who pursues significances and values is often compared unfavourably with the scientist who pursues atoms and electrons. The plain matter-of-fact person is disposed to think that the former is wandering amid shadow and illusion, whilst the latter is coming to grips with reality. I want therefore to give an illustration which will show that unless we pay attention to significances as well as to physical entities we may miss the essential part of experience.

Let us suppose that on November 11th a visitor from another planet comes to the Earth in order to observe scientifically the phenomena occurring here. He is especially interested in the phenomena of sound, and at the moment he is occupied in observing the rise and fall of the roar of traffic in a great city. Suddenly the noise ceases, and for the space of two minutes there is the utmost stillness; then the roar begins again. Our visitor, seeking a scientific explanation of this, may perhaps recall that on another occasion he witnessed an apparently analogous phenomenon in the kindred study of light. It was full daylight, but there came a quick falling of darkness which lasted about two minutes, after which the light came back again. The latter occurrence (a total eclipse of the Sun) has a well-known scientific explanation and can indeed be predicted many years in advance. I am assuming that the visitor is a competent scientist; and though he might at first be misled by the resemblance, he would soon find that the cessation of sound was a much more complicated phenomenon than the cessation of light. But there is nothing to suggest that it was outside the operation of the same kind natural forces. There was no supernatural hushing of sound. The noise ceased because the traffic stopped; each car stopped because a brake applied the necessary friction; the brake was worked mechanically by a pedal; the pedal by a foot; the foot by a muscle; the muscle by mechanical or electrical impulses travelling along a nerve. The stranger may well believe that each motion has its physical antecedent cause which can be carried back as far as we please; and if the prediction of the two-minute silence on Armistice Day is not predictable like an eclipse of the sun it is only because of the difficulty of dealing with the configurations of millions of particles instead of with a configuration of three astronomical bodies.

I do not myself think that the intermission of sound was predictable solely by physical laws. It might have been foreseen some days in advance if the visitor had access to the thoughts floating in human minds, but not from any study however detailed of the physical constituents of human brains. I think I am right in saying that within the last two years there has been a change in scientific ideas which makes this more likely than the old deterministic view. But here I am going to grant our visitor his claim; to concede that even human actions are predictable by a – possibly enlarged – scheme of physical law. What then? Shall we let our visitor go away convinced that he has got to the bottom of the phenomenon of Armistice Day? He understands perfectly why there is a two-minute silence; it is a natural and calculable result of the motion of a number of atoms and electrons following Maxwell's equations and the laws of conservation. It differs only from a similar optical event of a two-minute eclipse in being more complicated. Our visitor has apprehended the reality underlying the silence, so far as reality is a matter of atoms and electrons. But he is unaware that the silence has also a significance.

Often the best way to turn aside an attack is to concede it. The more complete the scientific explanation of the silence the more irrelevant that explanation becomes to our experience. When we assert that God is real, we are not restricted to a comparison with the reality of atoms and electrons. If God is as real as the shadow of the Great War on Armistice Day, need we seek further reason for making a place for God in our thoughts and lives? We shall not be concerned if the scientific explorer reports that he is perfectly satisfied that he has got to the bottom of things without having come across either.

VII

Assurance of the revelation of God rather than of the existence of God is demanded

We want an assurance that the soul in reaching out to the unseen world is not following an illusion.

We want security that faith, and worship, and above all love, directed towards the environment of the spirit are not spent in vain. It is not sufficient to be told that it is good for us to believe this, that it will make better men and women of us. We do not want a religion that deceives us for our own good. There is a crucial question here; but before we can answer it, we must frame it.

The heart of the question is commonly put in the form "Does God really exist?" It is difficult to set aside this question without being suspected of quibbling. But I venture to put it aside because it raises so many unprofitable side issues, and at the end it scarcely reaches deep enough into religious experience. Among leading scientists to-day I think about half assert that the æther exists and the other half deny its existence; but as a matter of fact both parties mean exactly the same thing, and are divided only by words. Ninety-nine people out of a hundred have not seriously considered what they mean by the term "exist" nor how a thing qualifies itself to be labelled real. A late colleague of mine, Dr. MacTaggart, wrote a two-volume treatise on "The Nature of Existence" which may possibly contain light on the problem, though I confess I doubt it. Theological or anti-theological argument to prove or disprove the existence of a deity seems to me to occupy itself largely with skating among the difficulties caused by our making a fetish of this word. It is all so irrelevant to the assurance for which we hunger. In the case

of our human friends we take their existence for
granted, not caring whether it is proven or not.
Our relationship is such that we could read
philosophical arguments designed to prove the
non-existence of each other, and perhaps even be
convinced by them – and then laugh together over
so odd a conclusion. I think that it is something of
the same kind of security we should seek in our
relationship with God. The most flawless proof of
the existence of God is no substitute for it; and if
we have that relationship the most convincing
disproof is turned harmlessly aside. If I may say it
with reverence, the soul and God laugh together
over so odd a conclusion.

For this reason I do not attach great importance to
the academic type of argument between atheism
and deism. At the most it may lead to a belief that
behind the workings of the physical universe there
is need to postulate a universal creative spirit, or it
may be content with the admission that such an
inference is not excluded. But there is little in this
that can affect our human outlook. It scarcely
amounts even to a personification of Nature; God is
conceived as an all-pervading force, which for
rather academic reasons is not to be counted
among forces belonging to physics. Nor does this
pantheism awake in us feelings essentially
different from those inspired by the physical world
– the majesty of the infinitely great, the marvel of
the infinitely little. The same feeling of wonder and
humility which we feel in the contemplation of the

stars and nebulae is offered as before; only a new name is written up over the altar. Religion does not depend on the substitution of the word "God" for the word "Nature."

The crucial point for us is not a conviction of the existence of a supreme God but a conviction of the revelation of a supreme God. I will not speak here of the revelation in a life that was lived nineteen hundred years ago, for that perhaps is more closely connected with the historical feeling which, equally with the scientific feeling, claims a place in most men's outlook. I confine myself to the revelation implied in the indwelling of the divine spirit in the mind of man.

It is probably true that the recent changes of scientific thought remove some of the obstacles to a reconciliation of religion with science; but this must be carefully distinguished from any proposal to base religion on scientific discovery. For my own part I am wholly opposed to any such attempt. Briefly the position is this. We have learnt that the exploration of the external world by the methods of physical science leads not to a concrete reality but to a shadow world of symbols, beneath which those methods are unadapted for penetrating. Feeling that there must be more behind, we return to our starting point in human consciousness – the one centre where more might become known. There we find other stirrings, other revelations (true or false) than those conditioned by the world of symbols. Are not these too of significance? We can only answer according to our conviction, for here reasoning fails us altogether. Reasoning leads us from premises to conclusion; it cannot start without premises. The premises for our reasoning about the visible universe, as well as for our reasoning about the unseen world, are in the self-knowledge of mind. Obviously we cannot trust every whim and fancy of the mind as though it were indisputable revelation; we can and must believe that we have an inner sense of values which guides us as to what is to be heeded, otherwise we cannot start on our survey even of the physical world. Consciousness alone can determine the validity of its convictions. "There shines no light save its own light to show itself unto itself."

The study of the visible universe may be said to start with a determination to use our eyes. At the very beginning there is something which might be described as an act of faith – a belief that what our eyes have to show us is significant. I think it can be maintained that an analogous determination that the mystic recognises another faculty of consciousness and accepts as significant the vista of a world outside space and time that it reveals. But if they start alike, the two outlets from consciousness are followed up by very different methods; and here we meet with a scientific criticism which seems to have considerable justification. It would be wrong to condemn alleged knowledge of the unseen world because it is unable

to follow the lines of deduction laid down by
science as appropriate to the seen world; but
inevitably the two kinds of knowledge are
compared, and I think the challenge to a
comparison does not come wholly from scientists.
Reduced to precise terms, shorn of words that
sound inspiring but mean nothing definite, is our
scheme of knowledge of what lies in the unseen
world, and of its mode of contact with us, at all to
be compared with our knowledge (imperfect as it
is) of the physical world and its interaction with
us? Can we be surprised that the student of
physical science ranks it rather with the vague
unchecked conjectures in his own subject, on which
he feels it his duty to frown? It may be that, in
admitting that the comparison is unfavourable, I
am doing an injustice to the progress made by
systematic theologians and philosophers; but at
any rate their defence had better be in other hands
than mine.

Although I am rather in sympathy with this
criticism of theology, I am not ready to press it to
an extreme. In this lecture I have for the most part
identified science with physical science. This is not
solely because it is the only side for which I can
properly speak, but because it is generally agreed
that physical science comes nearest to that
complete system of exact knowledge which all
sciences have before them as an ideal. Some fall far
short of it. The physicist who inveighs against the
lack of coherence and the indefiniteness of

theological theories, will probably speak not much less harshly of the theories of biology and psychology. They also fail to come up to his standard of methodology. On the other side of him stands an even superior being – the pure mathematician – who has no high opinion of the methods of deduction used in physics, and does not hide his disapproval of the laxity of what is accepted as proof in physical science. And yet somehow knowledge grows in all these branches. Wherever a way opens we are impelled to seek by the only methods that can be devised for that particular opening, not over-rating the security of our finding, but conscious that in this activity of mind we are obeying the light that is in our nature.

VIII

In everyday life (both material and spiritual) scientific analysis supplements but must not supplant a familiar outlook

I have said that the science of the visible universe starts with a determination to use our eyes; but that does not mean that the primary use of the eye is for advancing science. If in a community of the blind one man suddenly received the gift of sight, he would have much to tell which would not be at all scientific. Can we imagine him attempting to convey to his neighbours the significance of the new revelation by talking about the so-called physical "realities"? We know through science that

the differences of colour in the external world –
red, green, blue – are simply differences of
electromagnetic wave-length; and the existence of
colour-blindness shows how subjective the effects
of the waves on our senses may be. But to the man
who has received the revelation of sight the
significant fact is not so much the truth about
wave-length as the amazing transformation into a
world of colour under the vivifying power of the
mind. I need not stress the bearing of this when
the eye of the soul is opened to an apprehension of
the unseen world. The need for expression will not
satisfy itself in preaching a scientific sermon. In
the world, seen or unseen, there is place for
adventure as well as for triangulation. It is right
that we should, as far as may be, systematise and
criticise the inferences that may be drawn as to the
nature of the spiritual world beyond our
consciousness; but whatever its abstract frame
may be, it is transformed into a different
significance when it comes into relation with our
consciousness—even as the skeleton frame of
scientific truth is transformed into the colour and
activity and substance of our familiar
environment.

It seems right at this point to say a few words in
relation to the question of a Personal God. I
suppose every serious thinker is rather afraid of
this term which might seem to imply that he
pictures the deity on a throne in the sky after the
manner of medieval painters. There is a tendency

to substitute such terms as "omnipotent force" or even a "fourth dimension". If the idea is merely to find a wording which shall be sufficiently vague, it is somewhat unsuitable for the scientist to whom the words "force" and "dimension" convey something entirely precise and defined. On the other hand, my impression of psychology suggests that the word "person" might be considered vague enough as it stands. But leaving aside verbal questions I believe that the thought that lies behind this reaction is unsound. It is, I think, of the very essence of the unseen world that the conception of personality should dominate it. Force, energy, dimensions belong to the world of symbols; it is out of such conceptions that we have built up the external world of physics. What other conceptions have we? After exhausting physical methods we returned to the inmost recesses of conscious-ness, to the voice that proclaims our personality; and from there we entered on a new outlook. We have to build the spiritual world out of symbols taken from our own personality, as we build the scientific world out of the symbols of the mathematician. I think therefore we are not wrong in embodying the significance of the spiritual world to ourselves in the feeling of a personal relationship, for our whole approach to it is bound up with those aspects of consciousness in which personality is centred.

It is difficult to adjust the claims of naïve impressionism and scientific analysis of the spiritual realm without seeming to disparage one or the other; but I think it only requires the same commonsense that we apply to the affairs of ordinary life. Science has an important part to play in our everyday existence, and there is far too much neglect of science; but its intention is to supplement not to supplant the familiar outlook. The biochemist can teach us about the proteins and carbohydrates that make up a suitable diet and we may profit by his knowledge; but it is not fitting that a meal should be looked upon entirely from the standpoint of absorbing a specified quantity of calories and food-values. It would be still more absurd for a man to refuse food, because he was sceptical as to the certainty of the theories of bio-chemists. Likewise it is well that there should be some to advise us whether our spiritual bread contains the right kind of vitamins; but for the most part it is the object of our teaching and our meetings to stimulate the spiritual appetite rather than to conduct this kind of research.

If the kind of controversy which so often springs up between modernism and traditionalism in religion were applied to more commonplace affairs of life we might see some strange results. Would it be altogether unfair to imagine something like the following series of letters in our correspondence columns? It arises, let us say, from a passage in an obituary notice which mentions that the deceased had loved to watch the sunsets from his peaceful country home. *A*. writes deploring that in this progressive age few of the younger generation ever notice a sunset; perhaps this is due the pernicious influence of the teaching of Copernicus who maintains that the sun is really stationary. This rouses *B.* to reply that nowadays every reasonable person accepts Copernicus's doctrine. *C.* is positive that he has many times seen the sun set, and Copernicus must be wrong. *D.* calls for a restatement of belief, so that we may know just how much modern science has left of the sunset, and appreciate the remnant without disloyalty to truth. *E.* (perhaps significantly my own initial) in a misguided effort for peace points out that on the most modern scientific theory there is no absolute distinction between the heavens revolving round the earth and the earth revolving under the heavens; both parties are (relatively) right. *F.* regards this as a most dangerous sophistry, which insinuates that there is no essential difference between truth and untruth. *G.* thinks that we ought now to admit frankly that the revolution of the heavens is a myth; nevertheless such myths have still a practical teaching for us at the present day. *H.* produces an obscure passage in the Almagest, which he interprets as showing that the philosophy of the ancients was not really opposed to the Copernican view. And so it goes on. And the simple reader feels himself in an age of disquiet, insecurity and dissension, all because it is forgotten that what the deceased man looked out for each evening was an experience and not a creed.

IX

The spirit of Seeking in science and in religion

In its early days our Society owed much to a people who called themselves Seekers; they joined us in great numbers and were prominent in the spread of Quakerism. It is a name which must appeal strongly to the scientific temperament. The name has died out, but I think that the spirit of seeking is still the prevailing one in our faith, which for that reason is not embodied in any creed or formula. It is perhaps difficult sufficiently to emphasise Seeking without disparaging its correlative Finding. But I must risk this, for Finding has a clamorous voice that proclaims its own importance; it is definite and assured, some-thing that we can take hold of – that is what we all want, or think we want. Yet how transitory it proves. The finding of one generation will not serve for the next. It tarnishes rapidly except it be preserved with an ever-renewed spirit of seeking. It is the same too in science. How easy in a popular lecture to tell of the findings, the new discoveries which will be amended, contradicted, superseded in the next fifty years! How difficult to convey the scientific spirit of seeking which fulfils itself in this tortuous course of progress towards truth! You will understand the true spirit neither of science nor of religion unless seeking is placed in the forefront.

Religious creeds are a great obstacle to any full sympathy between the outlook of the scientist and the outlook which religion is so often supposed to require. I recognise that the practice of a religious community cannot be regulated solely in the interests of its scientifically-minded members and therefore I would not go so far as to urge that no kind of defence of creeds is possible. But I think it may be said that Quakerism in dispensing with creeds holds out a hand to the scientist. The scientific objection is not merely to particular creeds which assert in outworn phraseology beliefs which are either no longer held or no longer convey inspiration to life. The spirit of seeking which animates us refuses to regard any kind of creed as its goal. It would be a shock to come across a university where it was the practice of the students to recite adherence to Newton's laws of motion, to Maxwell's equations and to the electro-magnetic theory of light. We should not deplore it the less if our own pet theory happened to be included, or if the list were brought up to date every few years. We should say that the students cannot possibly realise the intention of scientific training if they are taught to look on these results as things to be recited and subscribed to. Science may fall short of its ideal, and although the peril scarcely takes this extreme form, it is not always easy, particularly in popular science, to maintain our stand against creed and dogma. I would not be sorry to borrow for our scientific pronouncements the passage prefixed to the Advices of the Society of Friends in 1656 and repeated in the current General Advices:

"These things we do not lay upon you as a rule or form to walk by; but that all with a measure of the light, which is pure and holy, may be guided; and so in the light walking and abiding, these things may be fulfilled in the Spirit, not in the letter; for the letter killeth, but the Spirit giveth life."

Rejection of creed is not inconsistent with being possessed by a living belief. We have no creed in science, but we are not lukewarm in our beliefs. The belief is not that all the knowledge of the universe that we hold so enthusiastically will survive in the letter; but a sureness that we are on the road. If our so-called facts are changing shadows, they are shadows cast by the light of constant truth. So too in religion we are repelled by that confident theological doctrine which has settled for all generations just how the spiritual world is worked; but we need not turn aside from the measure of light that comes into our experience showing us a Way through the unseen world.

Religion for the conscientious seeker is not all a matter of doubt and self-questionings. There is a kind of sureness which is very different from cocksureness.

Lightning Source UK Ltd.
Milton Keynes UK
UKHW040603191119
353823UK00001B/69/P

9 780901 689818